JUL 2020

S0-CYF-763

my guide to the planets

Saturn

CHERRY LAKE PRESS

Published in the United States of America by Cherry Lake Publishing
Ann Arbor, Michigan
www.cherrylakepublishing.com

Reading Adviser: Marla Conn, MS, Ed, Literacy specialist, Read-Ability, Inc.
Book Designer: Jennifer Wahi
Illustrator: Jeff Bane

Photo Credits: ©Vadim Sadovski/Shutterstock, 5, 7, 17, 19, 21; ©Triff/Shutterstock, 9, 11, 13; ©Vagengeim/Shutterstock, 15; ©NASA images/Shutterstock, 23; Cover, 2-3, 12, 16, 22, Jeff Bane; Various vector images throughout courtesy of Shutterstock.com/

Copyright ©2020 by Cherry Lake Publishing
All rights reserved. No part of this book may be reproduced or utilized in any form or by any means without written permission from the publisher.

Library of Congress Cataloging-in-Publication Data

Names: Devera, Czeena, author. | Bane, Jeff, 1957- illustrator. | Devera, Czeena. My guide to the planets.
Title: Saturn / by Czeena Devera ; illustrated by Jeff Bane.
Description: Ann Arbor, Michigan : Cherry Lake Publishing, [2020] | Series: My guide to the planets | Includes index. | Audience: K-1.
Identifiers: LCCN 2019032909 (print) | LCCN 2019032910 (ebook) | ISBN 9781534158849 (hardcover) | ISBN 9781534161146 (paperback) | ISBN 9781534159990 (adobe pdf) | ISBN 9781534162297 (ebook)
Subjects: LCSH: Saturn (Planet)--Juvenile literature.
Classification: LCC QB671 .D48 2020 (print) | LCC QB671 (ebook) | DDC 523.46--dc23
LC record available at https://lccn.loc.gov/2019032909
LC ebook record available at https://lccn.loc.gov/2019032910

Printed in the United States of America
Corporate Graphics

table of contents

About Saturn 4

Glossary 24

Index 24

About the author: Czeena Devera grew up in the red-hot heat of Arizona surrounded by books. Her childhood bedroom had built-in bookshelves that were always full. She now lives in Michigan with an even bigger library of books.

About the illustrator: Jeff Bane and his two business partners own a studio along the American River in Folsom, California, home of the 1849 Gold Rush. When Jeff's not sketching or illustrating for clients, he's either swimming or kayaking in the river to relax.

About Saturn

I'm Saturn. I'm the sixth-closest planet to the Sun.

I'm the second-largest planet in the **solar system**.

I **orbit** around the Sun. It takes me 29 years to complete 1 orbit!

My days are very short. One day for me is only around 10 hours!

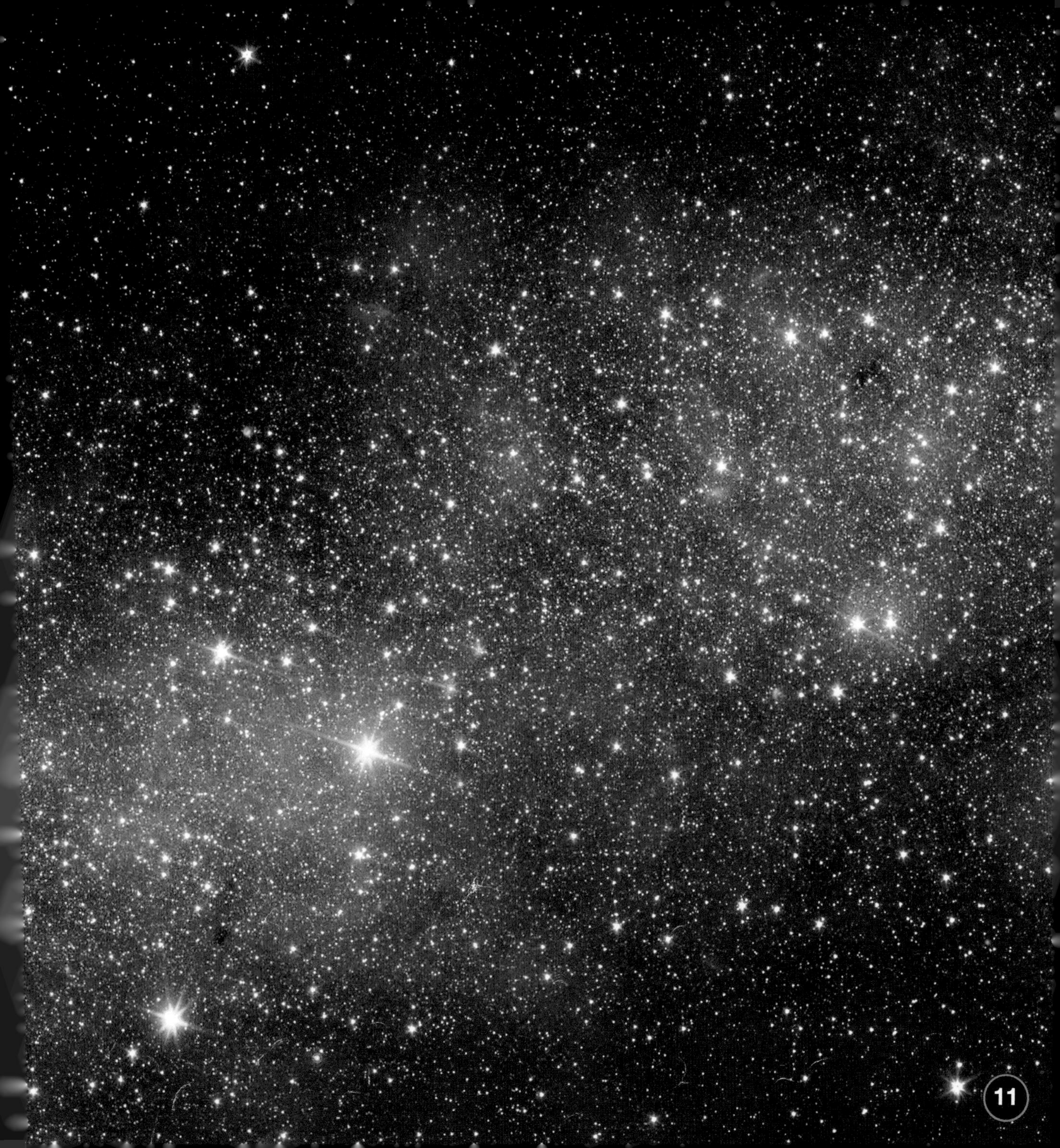

I am one of the four **gas giants**.

Gas giants are mostly made of gas. I don't have solid ground like Earth.

I am called the ringed planet. I have seven ring groups surrounding me.

My rings are made of ice and rocks. They can be easily seen with a **telescope**.

I have 82 moons. That's more than any other planet!

I am a **unique** planet. There are still many things **scientists** are discovering about me.

glossary & index

glossary

gas giants (GAS JYE-uhnts) planets that are made up mostly of gases and are not solid

orbit (OR-bit) to travel in a curved path around something

scientists (SYE-uhn-tists) people who study nature and the world we live in

solar system (SOH-lur SIS-tuhm) the sun and all the things that orbit around it, like planets

telescope (TEL-uh-skope) an instrument that makes distant objects seem larger and closer

unique (yoo-NEEK) the only one of its kind

index

Earth, 14

moons, 20

rings, 16, 18

gas giants, 12, 14

orbit, 8

Sun, 4, 8